PERENNIAL MATH

CONTEST PROBLEMS

2018-2019

Dr. Sylvia Dean was a gifted specialist and administrator for 26 years in Huntsville, Alabama. She created Perennial Math in 2011 upon completion of her doctoral degree. She has hosted many robotics, technology, and mathematics competitions throughout the years and has tirelessly worked on providing educational opportunities to students around the world. It has always been her ambition to leave students with the feeling that "math is cool."

These problems are challenging and students should know up front that scoring a 3 out of 6 on any of these tests is extraordinary and above average. Great problem-solvers learn persistence and appreciation for challenge. Good luck as you encourage your students to take the quest to become great mathletes.

Advanced

level

Grades 7-8

Name________________________________

PERENNIAL MATH

TEST #1

November 2018 Advanced Level

1) 20% of 30% of x is 210. What is x?	
2) How many paths exist from bottom corner X to top corner Y? You may only move to the left, up, or diagonally upward. 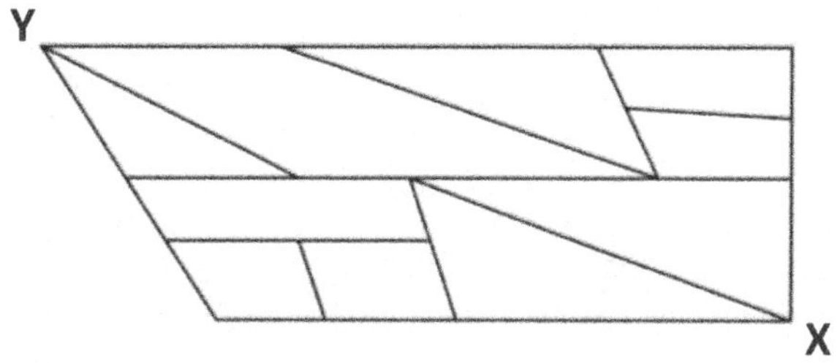	
3) Consider the sequence of numbers 2, 5, 8, 11, 14 . . . , in which each number increases by 3. What is the 99th number in the sequence?	
4) Big Spring Café has 19 tables. Some of the tables seat 6 people and some tables only seat 4 people. If a total of 90 people can be seated in the cafe, how many tables seat 6 people?	
5) The average age of the 6 people in room A is 40. The average age of the 4 people in room B is 25. If the people in the two rooms gather together, what is the average age of the 10 people?	
6) Of the 500 marbles in a large bowl, 80% are red and the rest are blue. How many of the red marbles must be removed from the bowl so that 75% of the remaining marbles are red?	

PERENNIAL MATH TEST #1

November 2018 Advanced Level Answers

1) 20% of 30% of x is 210. What is x?	3500
2) How many paths exist from bottom corner X to top corner Y? You may only move to the left, up, or diagonally upward. 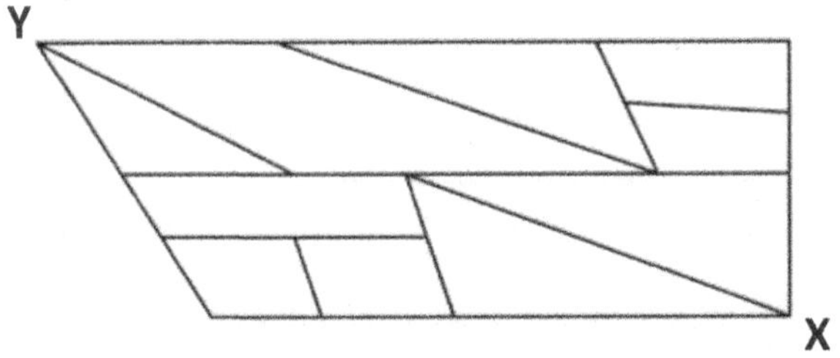	13
3) Consider the sequence of numbers 2, 5, 8, 11, 14 . . . , in which each number increases by 3. What is the 99th number in the sequence?	296
4) Big Spring Café has 19 tables. Some of the tables seat 6 people and some tables only seat 4 people. If a total of 90 people can be seated in the cafe, how many tables seat 6 people?	7
5) The average age of the 6 people in room A is 40. The average age of the 4 people in room B is 25. If the people in the two rooms gather together, what is the average age of the 10 people?	34
6) Of the 500 marbles in a large bowl, 80% are red and the rest are blue. How many of the red marbles must be removed from the bowl so that 75% of the remaining marbles are red?	100

PERENNIAL MATH TEST # 1

November 2018 Advanced Level Solutions

1) 20% of 30% of x is 210. What is x?

0.2 * .3 * x = 210

.06 * x = 210

x = 210 / 0.06

x = 3500

3500

2) There are 13 paths (as shown with the red path below) from X to Y. Remember you may only move to the left or up or diagonally upward.

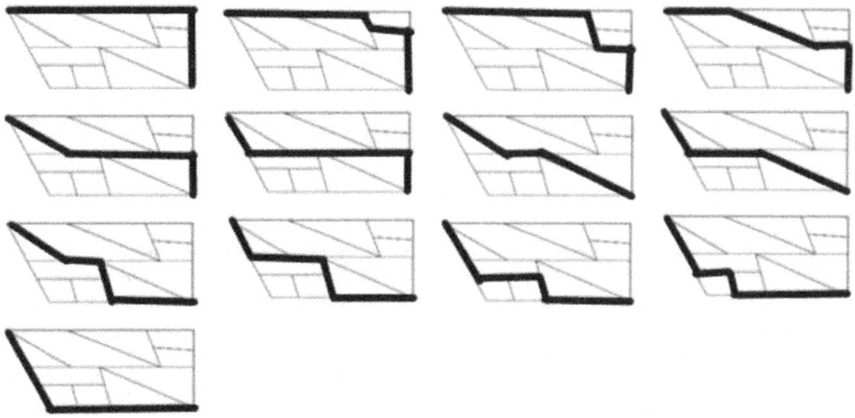

13

3) Consider the sequence of numbers 2, 5, 8, 11, 14 . . . , in which each number increases by 3. To find the 99th number in this sequence use the following formula: $a_n = a_1 + (n - 1)\, d$ where a_n is the term you are trying to find, a_1 is the first term, n is the number of terms, and d is the common difference between terms. So, $a_n = 2$, n = 99, and d = 3. Therefore, substitute these values into the formula:

$a_{99} = 2 + (99-1)3$

$a_{99} = 2 + (98)3$

$a_{99} = 2 + 294 = 296$

296

PERENNIAL MATH

TEST # 1

November 2018 Advanced Level Solutions

4) Create a table to find the number of tables that seat 6 people. Start somewhere in the middle for the number of each size of table. Note that for the 1st guess the total seats > 90, so decrease the number of 6 people tables until there are 90 total seats. There are 7 tables that will seat 6 people.

4 seat tables	10	11	12
6 seat tables	9	8	7
total tables (19)	19	19	19
total seats (90)	94	92	90

7

5) The average age of the 6 people in room A is 40. The average age of the 4 people in room B is 25. If the people in the two rooms gather together, what is the average age of the 10 people? To find the overall average age of the two rooms calculate the total of all of the ages and divide by the total number of people.

Room A: 6 people x 40 = 240

Room B: 4 people x 25 = 100

Therefore, total sum of the ages of 10 people is 240 + 100 = 340. The average of all 10 people is 340/10 = 34.

34

6) Of the 500 marbles in a large bowl, 80% are red and the rest are blue. How many of the red marbles must be removed from the bowl so that 75% of the remaining marbles are red? First calculate the number of red marbles at the start. The percent equation is Percent = (Given Quantity / Total Quantity) x 100. This can be rewritten as Quantity = (Percent x Total Quantity) / 100. So, at the start there are (80% x 500) /100 = 400 red marbles and the number of blue marbles is 500 - 400 = 100. Note the number of blue marbles will remain constant, but the total marbles will change as we remove the red marbles. We want to remove red marbles until the percentage of red marbles drops to 75%. Let the number of red marbles to be removed be x. The percent equation becomes:

$75 = \frac{(400-x)*100}{500-x}$ {multiply both sides by (500 – x)}

37500 - 75x = 40000 - 100x

25x = 2500

x = 100 red marbles removed so that 75% of the remaining marbles are red.

Or you can make a chart and remove the number of red marbles until you have 75% of the remaining marbles are red. So 400 - 300 = 100 red marbles to be removed.

# of red marbles	400	350	300
# of blue marbles	100	100	100
Total marbles	500	450	400
% red marbles	80	78	75

100

Name________________________________

PERENNIAL MATH TEST #2

December 2018 Advanced Level

1) 25 is 125% of what number?

2) Cindy chooses a number between 100 and 180. If the number leaves a remainder of 3 when it is divided 4, 5, and 6, what is her number?

3) Twelve girls went shopping and spent a total of $57. Julie spends more dollars than all of the other girls combined. Every girl spends a prime number of dollars including Julie. What is the least number of dollars that Julie could have spent?

4) What is the area of the shaded region in this diagram? Use $\pi = 3.14$.

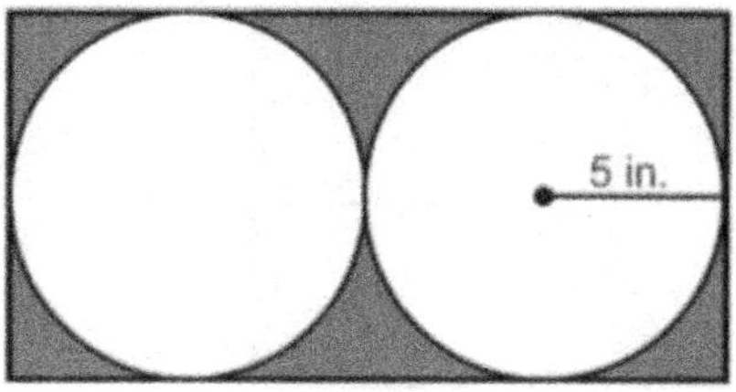

5) Given three boxes, there are three possible ways to choose a pair. If each pair weighs either 122, 125 or 127 pounds, what is the sum of the weights of all three boxes?

6) The stadium has 5000 seats. 2/5 of the seats are for season ticket holders. There were 1500 season ticket holders seated and some other seats already full. The stadium was 3/4 full. How many non-season ticket holders were seated?

PERENNIAL MATH

TEST # 2

December 2018 Advanced Level Answers

1) 25 is 125% of what number?	20
2) Cindy chooses a number between 100 and 180. If the number leaves a remainder of 3 when it is divided 4, 5, and 6, what is her number?	123
3) Twelve girls went shopping and spent a total of $57. Julie spends more dollars than all of the other girls combined. Every girl spends a prime number of dollars including Julie. What is the least number of dollars that Julie could have spent?	29
4) What is the area of the shaded region in this diagram? Use π =3.14. 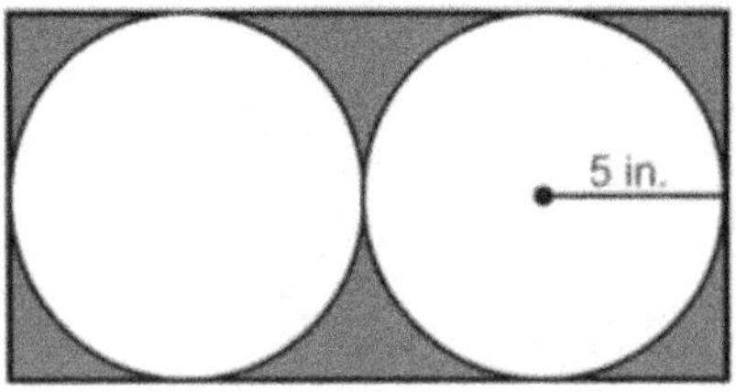	43
5) Given three boxes, there are three possible ways to choose a pair. If each pair weighs either 122, 125 or 127 pounds, what is the sum of the weights of all three boxes?	187
6) The stadium has 5000 seats. 2/5 of the seats are for season ticket holders. There were 1500 season ticket holders seated and some other seats already full. The stadium was 3/4 full. How many non-season ticket holders were seated?	2250

PERENNIAL MATH

TEST #2

December 2018 Advanced Level Solutions

1) 25 is 125% of what number?

25 = 1.25x

25/1.25 = x, x = 20.

20

2) Cindy chooses a number between 100 and 180. If the number leaves a remainder of 3 when it is divided 4, 5, and 6, what is her number? The least common multiple of 4, 5, and 6 is 60. Therefore, the only multiple between 100 and 180 is 120. Three more than 120 is 123. The number 123 would leave a remainder of 3 when it is divided by 4, 5, and 6.

123

3) Twelve girls went shopping and spent a total of $57. Julie spends more dollars than all of the other girls combined. Every girl spends a prime number of dollars including Julie. What is the least number of dollars that Julie could have spent? Make a list of the prime numbers and see what combinations give a total of $57. If the other eleven girls spent $2 each, that would be a total of $22. That would leave $35 that Julie spent. So, the list of prime numbers must be under 35:

2, 3, 5, 7, 11, 13, 17, 19, 23, 29, 31

Use trial and error:

-If Julie spends $31, the other girls combined could spend $26.

-If Julie spends $29, the other girls combined could spend $28.

-If Julie spends $23, the other girls combined could spend $34, which will not work because Julie has to spend more that the other girls combined.

So, $29 is the least amount that Julie could spend. Now check to see if the eleven girls could each spend a prime amount that adds up to $28. The other eleven girls could spend 2 + 2 + 2 + 2 + 2 + 2 + 2 + 3 + 3 + 3 + 5 = 28. Julie could spend the remaining $57 - $28 = $29.

29

PERENNIAL MATH

TEST #2

December 2018 Advanced Level Solutions

4) To find the area of the shaded region, find the area of the rectangle and subtract the area of both circles. Since the radius is 5in, then the length of the rectangle is 4 x 5in = 20in and the width is 2 x 5in = 10in. The area of the rectangle is 20in x 10in = $200in^2$. The area of the circle is 3.14 x $5in^2$ = $78.5in^2$. The area of 2 circles is $78.5in^2$ x 2 = $157in^2$. $200in^2$ - $157in^2$ = $43in^2$.

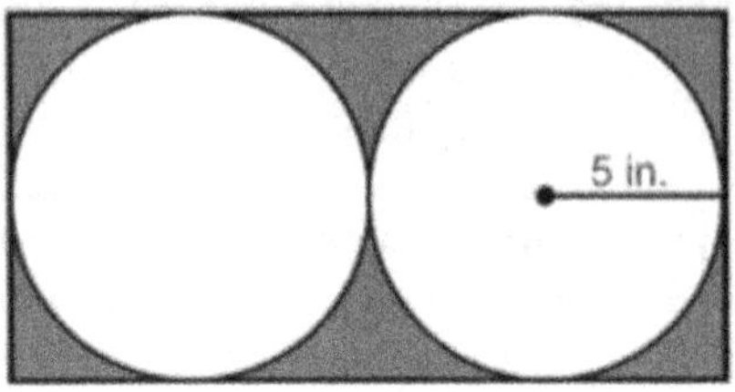

43

5) Given three boxes, there are three possible ways to choose a pair. If each pair weighs either 122, 125 or 127 pounds, what is the sum of the weights of all three boxes? Use algebra. Let A + B = 122, B + C = 125, and A + C = 127. Now just add all of the weights together and simplify.

$$\begin{array}{r} A + B = 122 \\ B + C = 125 \\ + \underline{A + C = 127} \\ 2A + 2B + 2C = 374 \\ A + B + C = 187 \end{array}$$

{Divide both sides in half }

187

6) Draw a picture. Draw the stadium in fifths and then separate these into fourths. The stadium is ¾ full which is 15/20. Each fourth holds 250 since 250 x 20 = 5000. There would be 15 sections filled and 5 sections empty. There are 1500 season ticket holders seated filling up 1500/250 = 6 of the 8 sections. This would leave 15 – 6 = 9 sections with non-season ticket holders for 9 x 250 = 2250 people.

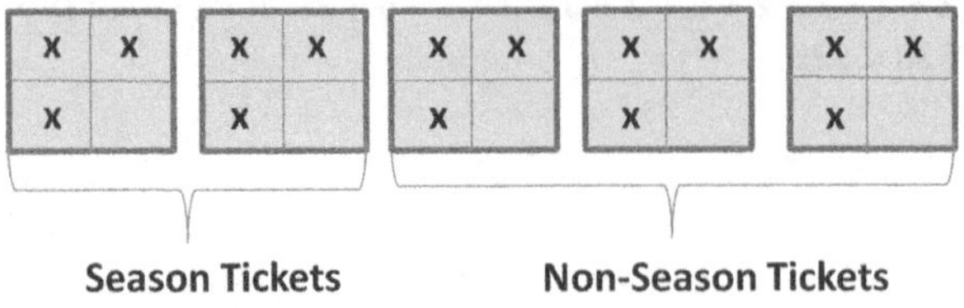

2250

Name________________________________

PERENNIAL MATH

January 2019 Advanced Level

TEST #3

1) What number goes in the square: 44 + □ = 72 - □	
2) A square and a triangle have equal perimeters. The lengths of the three sides of the triangle are 6.2 cm, 8.3 cm, and 9.5 cm. The area of the square is ____cm².	
3) In a game called Wham, three chirks and two dirks are worth 21 points, while two chirks and five dirks are worth 25 points. How many points are four chirks and three dirks worth?	
4) In the figure below, the different shapes are worth different amounts. Can you solve for the value of the circle? 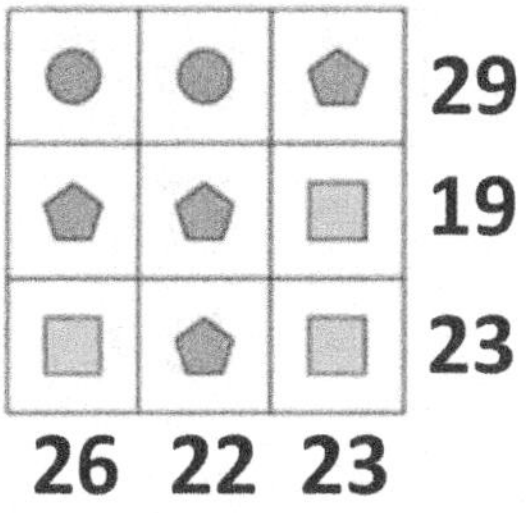	
5) How many square yards of carpet are required to cover a rectangular floor that is 18 feet long by 14.5 feet wide?	
6) What is the hundreds digit of 1 + 2 + 3 + ... + 98 + 99 + 100?	

PERENNIAL MATH — TEST #3

January 2019 Advanced Level Answers

1) What number goes in the square: 44 + ☐ = 72 - ☐	14
2) A square and a triangle have equal perimeters. The lengths of the three sides of the triangle are 6.2 cm, 8.3 cm, and 9.5 cm. The area of the square is ____cm^2.	36
3) In a game called Wham, three chirks and two dirks are worth 21 points, while two chirks and five dirks are worth 25 points. How many points are four chirks and three dirks worth?	29
4) In the figure below, the different shapes are worth different amounts. Can you solve for the value of the circle? 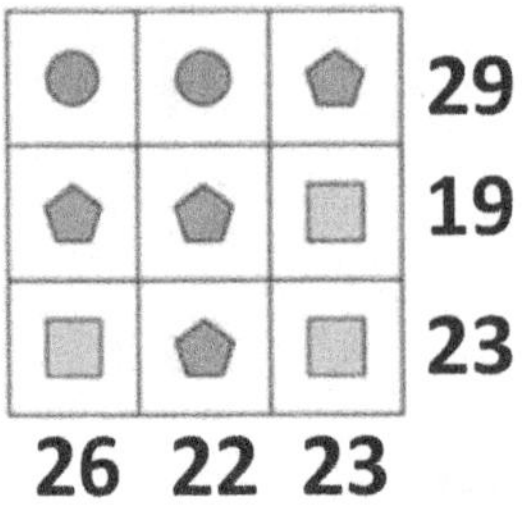	12
5) How many square yards of carpet are required to cover a rectangular floor that is 18 feet long by 14.5 feet wide?	29
6) What is the hundreds digit of 1 + 2 + 3 + ... + 98 + 99 + 100?	0

PERENNIAL MATH

TEST # 3

January 2019 Advanced Level Solutions

1) What number goes in the square: 44 + □ = 72 - □

Use algebra. $44 + x = 72 - x$

$$44 + x + x - 44 = 72 - x + x - 44 \quad \{\text{add } x - 44 \text{ to both sides}\}$$

$$2x = 28$$

$$x = 14$$

14

2) A square and a triangle have equal perimeters. The lengths of the three sides of the triangle are 6.2cm, 8.3cm, and 9.5cm. The perimeter of the triangle is 6.2 + 8.3 + 9.5 = 24cm. Since the square also has a perimeter of 24, its sides equal 24/4 or 6 cm. The area of the square is $36cm^2$.

36

3) In a game called Wham, three chirks and two dirks are worth 21 points, while two chirks and five dirks are worth 25 points. How many points are four chirks and three dirks worth? Use Algebra. Let C = chirks, D = dirks, and solve for 4C + 3D = ?

$$3C + 2D = 21 \quad \{\text{Equation 1}\}$$

$$-\ \underline{2C + 5D = 25}$$

$$C - 3D = -4$$

$$C = 3D - 4 \quad \{\text{Equation 2}\}$$

Now substitute this C into the Equation 1 and solve for D:

$$3C + 2D = 21$$

$$3(3D - 4) + 2D = 21$$

$$9D - 12 + 2D = 21$$

$$11D = 21 + 12$$

$$11D = 33$$

$$D = 3$$

Now substitute Equation 2 into 4C + 3D = ?

$$4(3(3) - 4) + 3(3) =$$

$$4(5) \quad + \quad 9 \quad = 29$$

29

PERENNIAL MATH

TEST #3

January 2019 Advanced Level Solutions

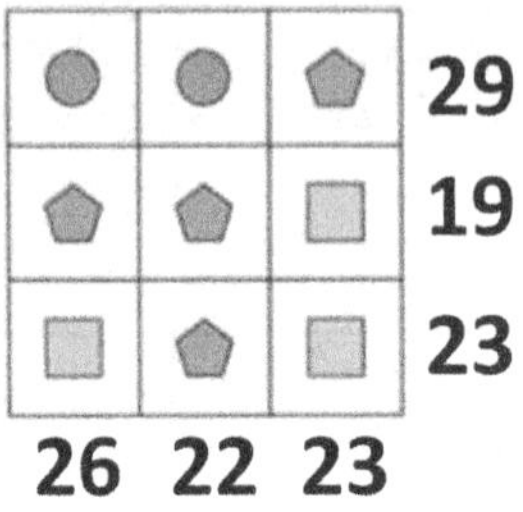

4) Use algebra. Let C = circle, P = pentagon, and S = square and solve for C. First solve for the value of S + P and then subtract that from C + S + P to solve for C.

2P + S = 19
P + 2S = 23
3P + 3S = 42

(3P + 3S) / 3 = 42/3 {Divide both sides by 3}
P + S = 14

Now subtract P + S = 14 from C + P + S = 26, C = 12.

12

5) How many square yards of carpet are required to cover a rectangular floor that is 18 feet long by 14.5 feet wide? First calculate the area in feet squared, Area = Length x Width = 18ft x 14.5ft = 261ft^2. Now convert that total to yards squared. Note that a squared yard = 3ft x 3ft = 9ft^2. So, number of square yards of carpet required is 261ft^2 ÷ 9 = 29yd^2.

29

6) A series such as 1 + 2 + 3 + ...+ 100 which has a constant difference between terms can be added quickly using a formula. The first term is a_1 is the first number in the sequence, and *n* is the number of terms. The sum of an arithmetic series is found by multiplying the number of terms time the average of the first and last terms.

$$S_n = \frac{n}{2}(a_1 + a_n)$$

S_{100}= 100/2(1 + 100)
S_{100}= 50(101) = 5<u>0</u>50

So, the number in the hundreds digit is 0.

0

Name________________________________

PERENNIAL MATH

TEST #4

February 2019 Advanced Level

1) If November 10, 2017 was on a Friday, what day of the week was February 28, 2018?	
2) The average of three different numbers is 30. If the lowest number is 17, what is the highest possible number of the remaining two numbers?	
3) If this figure was dropped in a bucket of red paint, what fraction of the smaller cubes would have red paint only on two sides? Answer must be in simplified form. 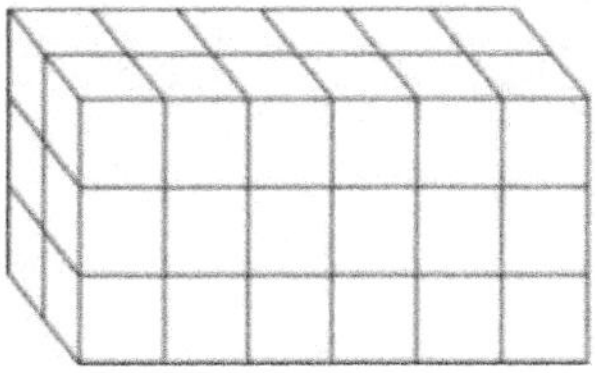	
4) The sum of three consecutive prime numbers is 211. What is the middle number?	
5) How many factors does 1260 have?	
6) How many distinct ways can the letters of the word "NUMBERS" be rearranged?	

PERENNIAL MATH TEST #4

February 2019 Advanced Level Answers

1) If November 10, 2017 was on a Friday, what day of the week was February 28, 2018?	Wednesday
2) The average of three different numbers is 30. If the lowest number is 17, what is the highest possible number of the remaining two numbers?	55
3) If this figure was dropped in a bucket of red paint, what fraction of the smaller cubes would have red paint only on two sides? Answer must be in simplified fraction form. 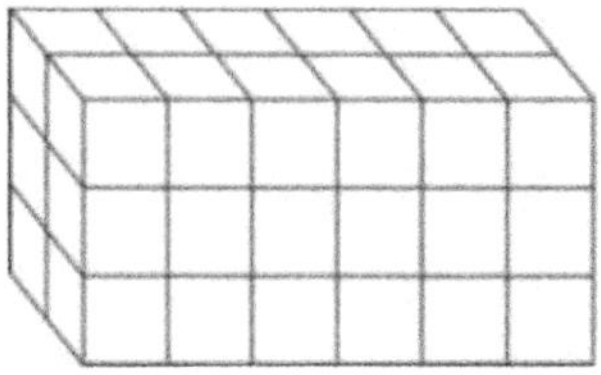	5/9
4) The sum of three consecutive prime numbers is 211. What is the middle number?	71
5) How many factors does 1260 have?	36
6) How many distinct ways can the letters of the word "NUMBERS" be rearranged?	5040

PERENNIAL MATH

TEST #4

February 2019 Advanced Level Solutions

1) If November 10, 2017 was on a Friday, what day of the week was February 28, 2018? Draw out a calendar and act it out. If November 10 is on a Friday, then the following Fridays would be November 17 and 24th, then December 1st would be on the following Friday because there are only 30 days in November. If December 1st is Friday, the following Fridays would be December 8, 15, 22, 29 and then January 1st would be on Monday since there are 31 days in December. Now if January 1st is a Monday, then the following Mondays would be January 8, 15, 22, 29 and February 1st would be on Thursday since there are 31 days in January. If February 1st is Thursday, then the following Thursdays would be February 8, 15, 22 and that would make February 28th on Wednesday.
Or, February 28th is 110 days (20 Nov. days + 31 Dec. days + 31 Jan. days + 28 Feb. days) from November 10th. 110 days / 7 days a week = 15 weeks and 5 days. Five days after Friday is a Wednesday.

Wednesday

2) The average of three different numbers is 30. The lowest number is 17. To find the highest possible number of the remaining two numbers, first figure out what the sum of the two remaining numbers has to be and then subtract 18 because it is the lowest possible number for the second number. The sum of the three numbers is 3 x 30 = 90. 90 – 17 = 73. 73 – 18 = 55.

55

3) If this figure was dropped in a bucket of red paint, what fraction of the smaller cubes would have red paint only on two sides? There are 2 x 3 x 6 = 36 blocks total. The blocks on the end corners and the very center of each side do not have 2 sides painted. There are 20 blocks that only have 2 sides painted. That is 20/36 = 5/9 of the blocks (in simplified form).

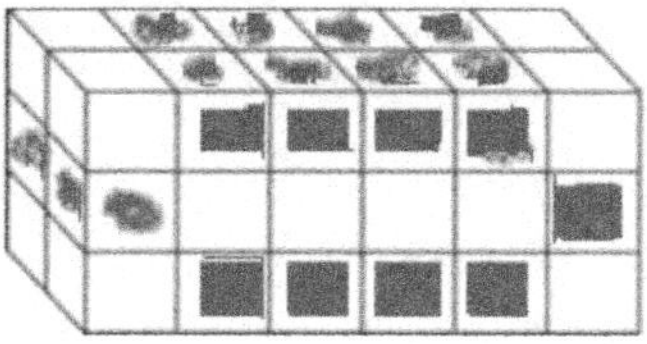

5/9

PERENNIAL MATH

TEST # 4

February 2019 Advanced Level Solutions

4) The sum of three consecutive prime numbers is 211. What is the middle number? First divide 211 by 3 to get an average and this will give you a starting point. Then list the prime numbers and guess and check. 211/3 = 70.3. The prime numbers around that include 67, 71, and 73. The sum of 67 + 71 + 73 = 211. The middle number of these consecutive prime numbers is 71.

71

5) How many factors does 1260 have? There are 18 factor pairs of 1260: 1 x 1260, 2 x 630, 3 x 420, 4 x 315, 5 x 252, 6 x 210, 7 x 180, 9 x 140, 10 x 126, 12 x 105, 14 x 90, 15 x 84, 18 x 70, 20 x 63, 21 x 60, 28 x 45, 30 x 42, and 35 x 36. Therefore 1260 has 36 factors.

36

6) How many distinct ways can the letters of the word "NUMBERS" be rearranged? The first letter can be any 1 of 7 letters. The second letter can now be any 1 of 6 letters. So the first two letters can be a combination of 7 × 6 letters = 42 arrangements. The third letter can be any 1 of 5 remaining letters. So the first three letters can be a combination of 7 × 6 × 5 letters = 210 arrangements. The fourth letter can now be any 1 of 4 remaining letters. So the first four letters can be a combination of 7 × 6 × 5 × 4 letters = 840 arrangements. The pattern would continue. So the first six letters can be a combination of 7 × 6 × 5 × 4 × 3 × 2 letters = 5040 arrangements. The final letter can now be only 1 letter because all of the others have been used. So the seven letters can be a combination of 5040 arrangements.

5040

Name________________________________

PERENNIAL MATH

TEST #1

January 2019 Advanced Level

1) If March 1, 2019 is on a Friday, then what day of the week will July 1, 2019 fall on?	
2) Given the sequence 1, 7, 13, 19…, what will be the first term that is greater than 2019?	
3) Target regions X, Y, and Z are worth different point values. The sum of X and Y is 25, the sum of Y and Z is 28, and the sum of X and Z is 27. How many points is Y worth? 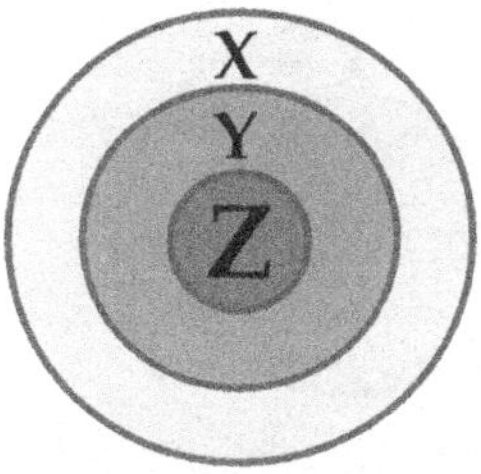	
4) Lance must make up a five-character password starting with 3 numbers (0-9) and followed by two letters (A-Z). Numbers and letters can be repeated within the password. How many possible password combinations can be made?	
5) There are many sets of three prime numbers whose sum is 33. What is the difference between the two sets with the greatest and least possible products?	
6) Pizza Planet sells pizzas with the choice of pepperoni, sausage and ham toppings. One day they sold 263 pizzas; 144 had pepperoni, 144 had sausage, 86 had pepperoni and ham, 64 had sausage but no ham, 70 had pepperoni and sausage, 39 had all three toppings and 20 had no toppings. How many pizzas were sold with only ham on them?	

PERENNIAL MATH

TEST # 1

January 2019 Advanced Level Answers

1) If March 1, 2019 is on a Friday, then what day of the week will July 1, 2019 fall on?	Monday
2) Given the sequence 1, 7, 13, 19..., what will be the first term that is greater than 2019?	2023
3) Target regions X, Y, and Z are worth different point values. The sum of X and Y is 25, the sum of Y and Z is 28, and the sum of X and Z is 27. How many points is Y worth? 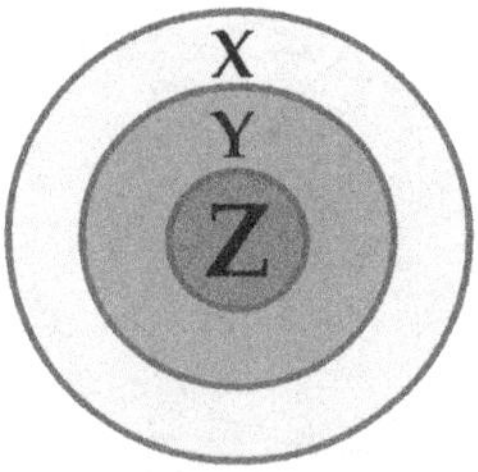	13
4) Lance must make up a five-character password starting with 3 numbers (0-9) and followed by two letters (A-Z). Numbers and letters can be repeated within the password. How many possible password combinations can be made?	676,000
5) There are many sets of three prime numbers whose sum is 33. What is the difference between the two sets with the greatest and least possible products?	1215
6) Pizza Planet sells pizzas with the choice of pepperoni, sausage and ham toppings. One day they sold 263 pizzas; 144 had pepperoni, 144 had sausage, 86 had pepperoni and ham, 64 had sausage but no ham, 70 had pepperoni and sausage, 39 had all three toppings and 20 had no toppings. How many pizzas were sold with only ham on them?	25

PERENNIAL MATH

TEST # 1

January 2019 Advanced Level Solutions

1) If March 1, 2019 is on a Friday, then what day of the week will July 1, 2019 fall on?
First calculate the number of days between March 1 and July 1. There are 31 days in March, 30 days April, 31 day in May, and 30 days in June for a total of 31 + 30 + 31 + 30 = 122. Now divide that by 7 (the number of day in a week), 122 / 7 = 17 with a remainder of 3. So, 3 days past Friday would be a Monday.

Monday

2) Given the sequence 1, 7, 13, 18…, what will be the first term that is greater than 2019? Note that the common difference between the terms is 6. To find out what term 2019 would be: 2019/6 = 336 plus a remainder.

$$a_n = a_1 + d(n-1).$$
$$a_{337} = 1 + 6(337-1)$$
$$a_{337} = 2017$$

The first term greater than 2019 would be the next term 2023, the 338th term.

2023

3) Target regions X, Y, and Z are worth different point values. The sum of X and Y is 25, the sum of Y and Z is 28, and the sum of X and Z is 27. How many points is Y worth? Use Algebra.

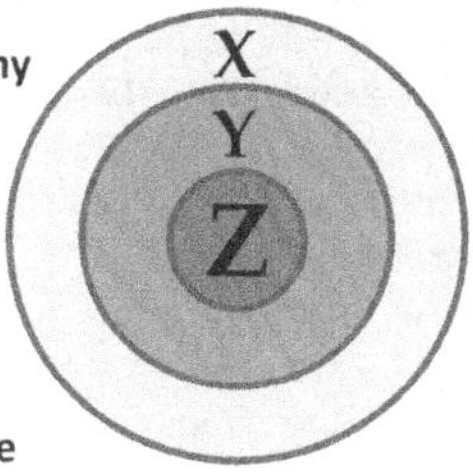

$$X + Y = 25$$
$$X + Z = 27$$
$$\underline{Y + Z = 28}$$
$$2X + 2Y + 2Z = 80 \quad \{\text{Now divide both sides by 2}\}$$
$$X + Y + Z = 40$$

Substitute the original equation X + Z = 27 back into X + Y + Z = 40, and solve for Y.

$$(27) + Y = 40$$
$$Y = 13$$

13

PERENNIAL MATH

TEST # 1

January 2019 Advanced Level Solutions

4) Lance must make up a five-character password starting with 3 numbers (0-9) and followed by two letters (A-Z). Numbers and letters can be repeated within the password. How many possible password combinations can be made? There are 10 possible choices for numbers and 26 for letters. Since repetition is allowed for all characters in the password, there are 10 x 10 x 10 x 26 x 26 = 676,000 possible combinations.

676,000

5) There are many sets of three prime numbers whose sum is 33. What is the difference between the two sets with the greatest and least possible products?
Make a list of the prime numbers 2, 3, 5, 7, 11, 13, 17, 19, 23, and 29. Now find the sets that have a sum of 33. The possible sets of three prime numbers whose sum is 33 are:
(2, 2, 29), (3, 7, 23), (3, 11, 19), (5, 5, 23), (5, 11, 17), (7, 13, 13), and (11, 11, 11).
The least possible product is 2 x 2 x 29 = 116. The greatest possible product is 11 x 11 x 11 = 1331.
Their difference is 1331 – 116 = 1215.

1215

6) Use a Venn Diagram. All regions and the outside region must add up to 263. The middle region that is intersected by all 3 rings is 39. Use this region to solve for the other regions. The 2 regions that are shared by pepperoni and ham must add up to 86. 86 – 39 = 47.
The two regions shared by pepperoni and sausage add up to add up to 70. 70 - 39 = 31.
The two regions in sausage that are not shared with ham add up to 64. 64 – 31 = 33.
The four regions for sausage must add up to 144.
144 – 31 – 39 – 64 = 41.
The four regions for pepperoni must add up to 144.
144 – 31 – 39 – 47 = 27.
Now find ham only by subtracting all other regions from total:
263 – (27 + 47 + 31 + 39 + 33 + 41 + 20) = 25
To check add all regions:
27 + 47 + 31 + 39 + 33 + 41 + 25 + 20 = 263 pizzas.

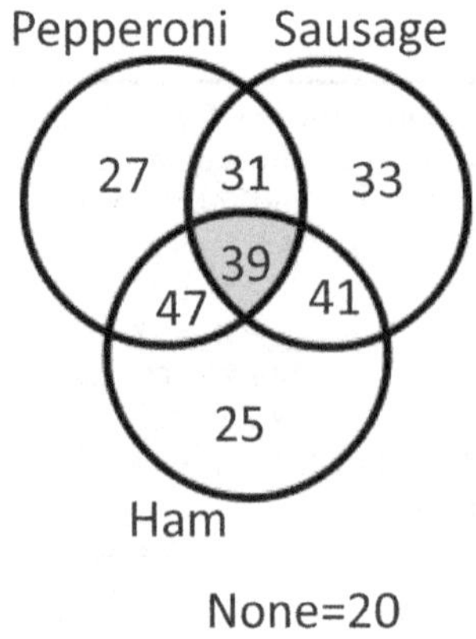

25

Name_______________________________

PERENNIAL MATH

February 2019 Advanced Level

TEST #2

1) Four times a number decreased by negative twelve is ninety-two. What is the number?	
2) One box of cereal contains 12% more cereal than another box. The larger box contains 3 more ounces of cereal than the smaller box. How many ounces of cereal are in the larger box?	
3) In a bakery, cupcakes and cookies have a different price. Two cupcakes and three cookies cost \$2.49. But three cupcakes and two cookies cost \$2.81. How much does one cupcake cost?	
4) If an 8 x 8 x 8 cube was painted red on all six faces and then cut up into 512 1 x 1 x 1 cubes, then how many of those small cubes would have an even number (including zero) of faces colored red? 	
5) The book shop had math and science books in the ratio of 2:5. At the end of the week 20% of each type of book was sold and 2240 total books remained unsold. How many total math books were there at the start of the week?	
6) Diamond has quarters, dimes, nickels and pennies in her bank. How many ways can she make 57 cents if she only has 10 of each coin?	

PERENNIAL MATH

TEST #2

February 2019 Advanced Level Answers

1) Four times a number decreased by negative twelve is ninety-two. What is the number?	20
2) One box of cereal contains 12% more cereal than another box. The larger box contains 3 more ounces of cereal than the smaller box. How many ounces of cereal are in the larger box?	28
3) In a bakery, cupcakes and cookies have a different price. Two cupcakes and three cookies cost \$2.49. But three cupcakes and two cookies cost \$2.81. How much does one cupcake cost?	.69
4) If an 8 x 8 x 8 cube was painted red on all six faces and then cut up into 512 1 x 1 x 1 cubes, then how many of those small cubes would have an even number (including zero) of faces colored red? 	288
5) The book shop had math and science books in the ratio of 2:5. At the end of the week 20% of each type of book was sold and 2240 total books remained unsold. How many total math books were there at the start of the week?	800
6) Diamond has quarters, dimes, nickels and pennies in her bank. How many ways can she make 57 cents if she only has 10 of each coin?	20

PERENNIAL MATH

TEST #2

February 2019 Advanced Level Solutions

1) Four times a number decreased by negative twelve is ninety-two. Use algebra to find the number.

$4x - (-12) = 92$

$4x + 12 = 80$

$4x = 80$

$x = 20$

20

2) One box of cereal contains 12% more cereal than another box. The larger box contains 3 more ounces of cereal than the smaller box. How many ounces of cereal are in the larger box? First solve for number of ounces in the smaller box and then add 3 ounces to get the larger box size. 3 ounces more than the smaller box is equivalent to 12%. So, 1 ounce is 4%. Therefore, the smaller box contains 100% ÷ 4 = 25 ounces. Since the larger box contains 3 more ounces so it contains 25 + 3 = 28 ounces.

28

3) In a bakery, cupcakes and cookies have a different price. Two cupcakes and three cookies cost $2.49. But three cupcakes and two cookies cost $2.81 cents. Use algebra to find the cost of one cupcake. Let x = cupcake and y = cookie.

$2x + 3y = 2.49$

$+ \underline{3x + 2y = 2.81}$ Equation 1

$5x + 5y = 5.30$ {divide both sides by 5}

$x + y = 1.06$ {now multiply both sides by 2 for convenience}

$2x + 2y = 2.12$ {subtract this equation from Equation 1 to solve for x}

$3x + 2y = 2.81$

$- \underline{2x + 2y = 2.12}$

$x = .69$ One cupcake cost 69 cents.

.69

PERENNIAL MATH TEST #2

February 2019 Advanced Level Solutions

4) To find the number of cubes with an even number of painted sides, first determine the number of cubes with an odd number of painted sides and subtract that from the total number of smaller cubes. There are a total of 8 x 8 x 8 = 512 small (1 x 1 x 1) cubes in this larger cube. The 8 corner cubes have an odd number of painted sides. Also the interior or 6 x 6 = 36 cubes of all 6 outer sides have an odd number of painted sides. That is 6 x 36 = 216 cubes. 216 + 8 = 224 total cubes with an odd number of painted sides. Therefore 512 - 224 = 288 cubes with an even number of painted sides.

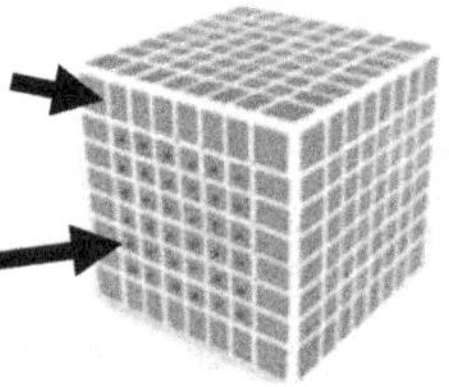

288

5) The book shop had math and science books in the ratio of 2:5. At the end of the week 20% of each type of book was sold and 2240 books remained unsold. How many math books were there at the start of the week? Since the books are in a ration of 2:5, there are 7 portions. Divide 2240/7 = 320 books per portion that were unsold. Math books make up 2 of those portions, so 320 x 2 = 640 math books unsold. 640 books is 80% of the math books. 640/x = 80/100. x = 800 total math books.

800

6) Diamond has quarters, dimes, nickels and pennies in her bank. How many ways can she make 57 cents if <u>she only has 10 of each coin</u>? Make an organized list. There are 20 ways to make 57 cents.

Q	D	N	P
2	0	1	2
2	0	0	7
1	3	0	2
1	2	2	2
1	2	1	7
1	1	4	2
1	1	3	7
1	0	6	2
1	0	5	7
0	5	1	2

Q	D	N	P
0	5	0	7
0	4	3	2
0	4	2	7
0	3	5	2
0	3	4	7
0	2	7	2
0	2	6	7
0	1	9	2
0	1	8	7
0	0	10	7

20

Name_______________________________

PERENNIAL MATH

March 2019 Advanced Level

TEST #3

1) What is 75% of 80% of 80?	
2) A bag contains 100 marbles: 45 blue, 33 orange, 12 green, and 10 red. If you are blindfolded, what is the least number of marbles you must take out to be sure that you have at least 14 marbles of the same color?	
3) What is the units digit of 7^{2019}?	
4) How many three-digit counting numbers do not contain any of the digits 1, 5, 7, or 8?	
5) Dominique received a 67, 73, 79, 72, 71 on her math tests. In order to pass the class, she needs an 80 average on her tests. She has two more tests left. What does she have to average on the last two tests in order to pass the class?	
6) What is the area of the shaded region? (Use $\pi = 3.14$) 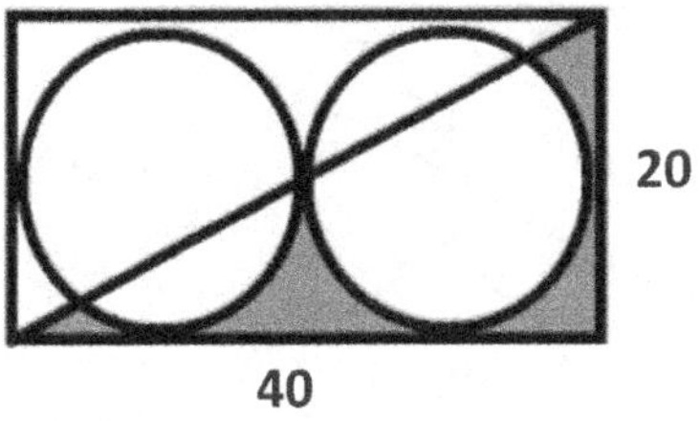	

PERENNIAL MATH

TEST # 3

March 2019 Advanced Level Answers

1) What is 75% of 80% of 80?	48
2) A bag contains 100 marbles: 45 blue, 33 orange, 12 green, and 10 red. If you are blindfolded, what is the least number of marbles you must take out to be sure that you have at least 14 marbles of the same color?	49
3) What is the units digit of 7^{2019}?	3
4) How many three-digit counting numbers do not contain any of the digits 1, 5, 7, or 8?	180
5) Dominique received a 67, 73, 79, 72, 71 on her math tests. In order to pass the class, she needs an 80 average on her tests. She has two more tests left. What does she have to average on the last two tests in order to pass the class?	99
6) What is the area of the shaded region? (Use $\pi = 3.14$) 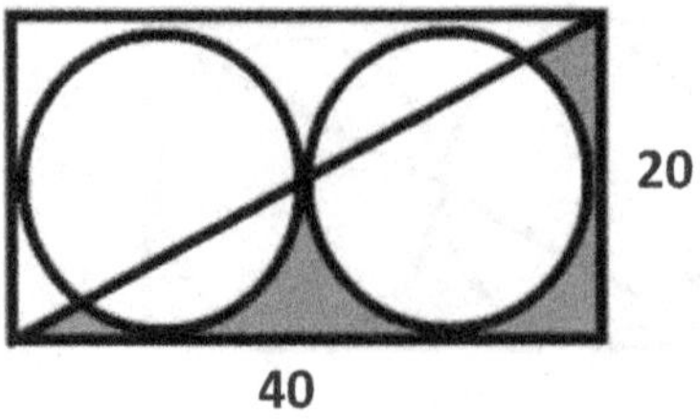	86

PERENNIAL MATH TEST #3

March 2019 Advanced Level Solutions

1) What is 75% of 80% of 80?

.75 x .8 x 80 = 48

48

2) A bag contains 100 marbles: 45 blue, 33 orange, 12 green, and 10 red. If you are blindfolded, what is the least number of marbles you must take out to be sure that you have at least 14 marbles of the same color? The first 10 marbles could be red, the next 12 could be green, then 13 could be blue and then the next 14 would definitely insure that you had 14 of the same color. So you must pick out 10 + 12 + 13 + 14 = 49 marbles in order to make sure you have at least 14 of the same color.

49

3) What is the units digit of 7^{2019}?

There is a pattern that repeats every fourth power:

$7^1 = \underline{7}$

$7^2 = 4\underline{9}$

$7^3 = 34\underline{3}$

$7^4 = 240\underline{1}$

$7^5 = 1680\underline{7}$

$7^6 = 11764\underline{9}$

...

...

To find the ones digit of 7^{2019} divide 2019th power by 4 and you get a remainder of 3. So the ones digit would be 3, which is the third digit in the pattern 7, 9, $\underline{3}$, 1.

3

PERENNIAL MATH

TEST # 3

March 2019 Advanced Level Solutions

4) How many three-digit counting numbers do not contain any of the digits 1, 5, 7, or 8?
That means that possible digits are 0, 2, 3, 4, 6, and 9. There are six possible digits for the units place. There are 6 possible digits for the tens place. However, there are only 5 possible digits for the hundreds digit because 0 cannot be in the hundreds place. Therefore, 6 x 6 x 5 = 180 possible combinations of the digits. There are 180 three-digit counting numbers that do not contain 1, 5, 7 or 8.

180

5) Dominique had a 67, 73, 79, 72, 71 on her math tests. In order to pass the class, she needs an 80 average on her tests. She has two more tests left. What does she have to average on the last two tests in order to pass the class? There are a total of seven tests since she has already taken 5 and has 2 tests left. In order to have an average of 80, the seven test scores must add up to 80 x 7 = 560. The current sum of her scores is 67 + 73 + 79 + 72 + 71 = 362. So, she must score 560 – 362 = 198 more points for the remaining two test in order to have an overall average of 80. Therefore, she must average 198/2 = 99 points on the last two tests.

99

6) The two parts of the circle that are dissected can be put together to form one complete circle. To find the area of the shaded region, first find the area of the triangle and subtract the area of just one of the circles.

Area of the triangle = ½ base x height
= (40)(20)/2
= 400

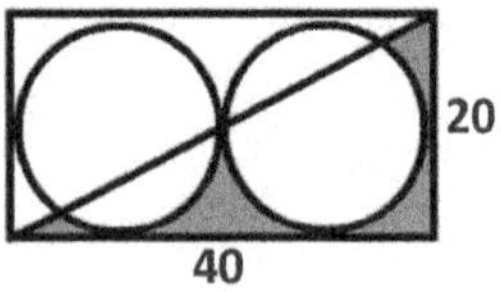

Radius of the circle = ½ x height
= 20/2 = 10
Area of circle = πr^2
Area of circle = $3.14 \times 10^2 = 314$

Area of shaded region is 400 - 314 = 86.

86

Name________________________________

PERENNIAL MATH

April 2019 Advanced Level

TEST #4

1) What is the positive square root of the number whose square is 256?

2) Four consecutive odd integers sum to 296. What is the least of the four integers?

3) Ted and Barry have 20 marbles in total. If Ted gives Barry 3 marbles, then Ted has 3 times as much as Barry. How many marbles does Ted have originally?

4) If Kevin walked for 45 minutes at a rate of 4mph and then ran for 30 minutes at a rate of 10mph, how many miles did he go at the end of one hour and 15 minutes?

5) What is the positive difference between the surface area in these two cubes? ____cm^2

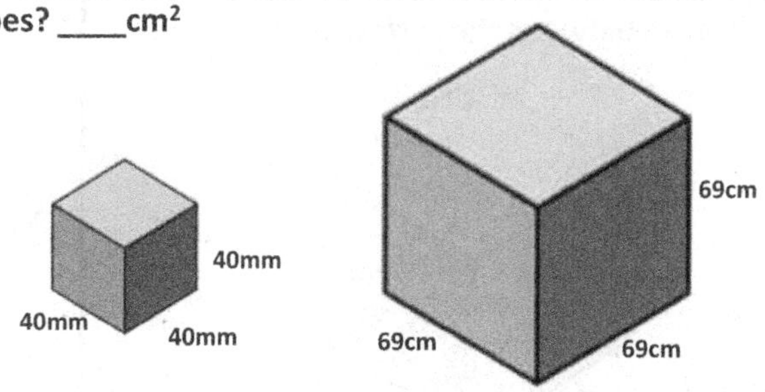

6) John mixes items to make various trail mixes. His Protein mix is 55% almond and 45% pistachio. His Energy mix is 30% almond and 70% raisin. His Hiking mix combines some of each of the Protein and Energy mixes. If the Hiking mix is 45% almond, how much of each kg of Hiking mix is Protein mix?

PERENNIAL MATH

TEST #4

April 2019 Advanced Level Answers

1) What is the positive square root of the number whose square is 256?	4
2) Four consecutive odd integers sum to 296. What is the least of the four integers?	71
3) Ted and Barry have 20 marbles in total. If Ted gives Barry 3 marbles, then Ted has 3 times as much as Barry. How many marbles does Ted have originally?	18
4) If Kevin walked for 45 minutes at a rate of 4mph and then ran for 30 minutes at a rate of 10mph, how many miles did he go at the end of one hour and 15 minutes?	8
5) What is the positive difference between the surface area in these two cubes? ____cm^2 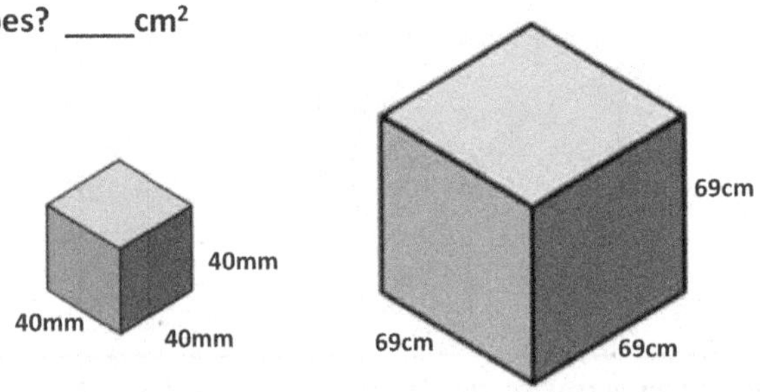	28,470
6) John mixes items to make various trail mixes. His Protein mix is 55% almond and 45% pistachio. His Energy mix is 30% almond and 70% raisin. His Hiking mix combines some of each of the Protein and Energy mixes. If the Hiking mix is 45% almond, how much of each kg of Hiking mix is Protein mix?	600

PERENNIAL MATH

TEST #4

April 2019 Advanced Level Solutions

1) What is the positive square root of the number whose square is 256? The number 256 is the square of 16. Therefore, the positive square root of that number (16) is 4.

4

2) Four consecutive odd integers sum to 296. What is the least of the four integers? First find the average. 296 divided by 4 = 74. Since 74 is the average, it would be the middle of all of the other 4 numbers. The four consecutive odd numbers who sum to 296 are 71 + 73 + 75 + 77. The smallest number is 71.

71

3) Ted and Barry have 20 marbles in total. If Ted gives Barry 3 marbles, then Ted has 3 times as much as Barry. How many marbles does Ted have originally?
In order for Ted to have three times as much as Barry, Barry will have a 3:1 ratio. This will mean he will have 15 marbles and Ted will have 5. Since he gave Ted 3, then Barry started with 15 + 3 = 18 marbles.

18

PERENNIAL MATH TEST #4

April 2019 Advanced Level Solutions

4) If Kevin walked for 45 minutes at a rate of 4mph and then ran for 30 minutes at a rate of 10mph, how many miles did he go at the end of one hour and 15 minutes? First convert the time into a decimal and multiply to find the number of miles traveled. 45 minutes is ¾ of an hour or .75 and 30 minutes is ½ hour or .5. Kevin walked at a rate of 4 x .75 = 3 miles. He ran at a rate of 10 x .5 = 5 miles. So he traveled 3 miles + 5 miles = 8 miles in the hour and 15 minutes.

8

5) The answer must be in cm^2 so convert the cube measurements to centimeters. Now solve for the surface area of each cube, and subtract to find the difference. 40mm = 4cm. Therefore, the surface area of the first cube is 6 sides × $16cm^2$ = $96cm^2$. The area of the second cube is 6 sides × $4761cm^2$ = $28{,}566cm^2$. Subtract the smallest from the largest to get the positive difference between the surface area of the two cubes, 28,566 cm^2 - $96cm^2$ = $28{,}470cm^2$.

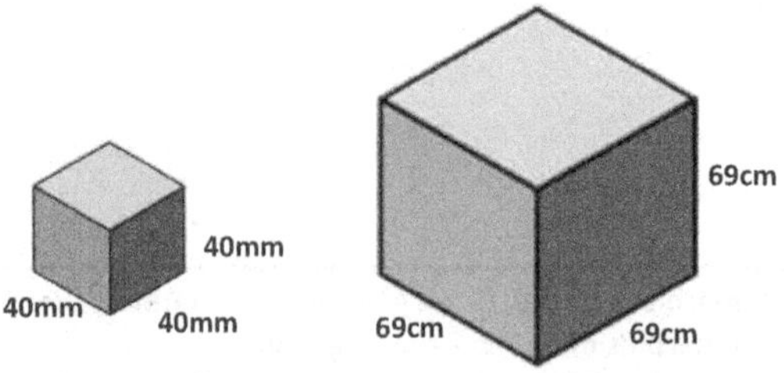

28,470

6) John mixes items to make various trail mixes. His Protein mix is 55% almond and 45% pistachio. His Energy mix is 30% almond and 70% raisin. His Hiking mix combines some of each of the Protein and Energy mixes. If the Hiking mix is 45% almond, how much of each kg of Hiking mix is Protein mix? This is a weighted average problem. If John mixed 500g of each mix, then his concentration of almonds would be (55% + 30%)/2 = 42.5%. So, this indicates that John needs a higher concentration of Protein mix. Use algebra to solve. Let C stand for concentration and V for volume.

C_1V_1 = the concentration times the volume of the Protein mix

C_2V_2 = the concentration times the volume of the Energy mix

C_3V_3 = the concentration times the volume of the Hiking mix

$$C_1V_1 + C_2V_2 = C_3V_3$$

$$55\%(V_1) + 30\%(V_2) = 45\%(1000g)$$

To solve V_1, express V_2 in terms of V_1, so $V_2 = 1000 - V_1$.

$$55(V_1) + 30(1000 - V_1) = 45(1000)$$

$$55V_1 + 30{,}000 - 30V_1 = 45{,}000$$

$$25V_1 = 15{,}000$$

$$V_1 = 600g$$

600

Name________________________________

PERENNIAL MATH

TEST #1

November 2019 Advanced Level

1) Solve for x. $2x - 5 = 13$.

2) The local pet store sells one cat for every 3 dogs. The shop also sells 4 cats for every 6 birds. How many dogs does the store sell for every bird?

3) Charmin ordered a 4 ft. long submarine sandwich to bring to the picnic. She made 15 cuts equally spaced along the length. How many inches long was each piece?

4)) A fence that costs $4.50 per foot is built in the dimensions shown below. How much did it cost to built the fence?

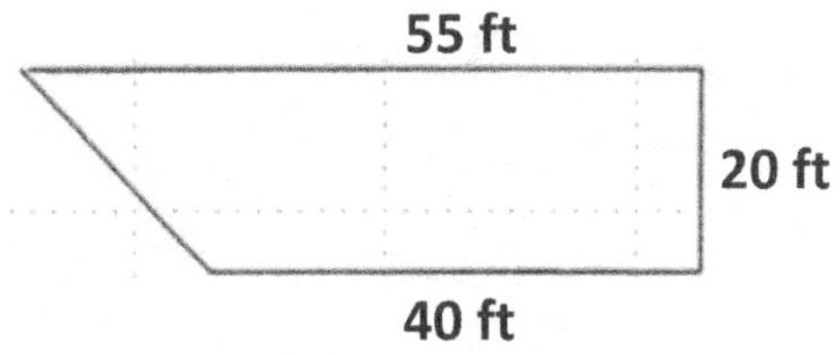

6) Jackson Middle arranged for 75 students to go to the theater to see a play. Some went in cars and some on a bus. Afterwards, the same bus and cars brought them back. Going to the play 51 students rode on the bus and 4 students rode in each car. Coming back 5 students rode in each car. How many students came back by bus?

5) A club of 10 people want to choose a president, vice-president, and treasurer. How many ways can they select them from the 10 people?

PERENNIAL MATH

TEST #1

November 2019 Advanced Level Answers

1) Solve for x. $2x - 5 = 13$.

9

2) The local pet store sells one cat for every 3 dogs. The shop also sells 4 cats for every 6 birds. How many dogs does the store sell for every bird?

2

3) Charmin ordered a 4 ft. long submarine sandwich to bring to the picnic. She made 15 cuts equally spaced along the length. How many inches long was each piece?

3

4) A fence that costs $4.50 per foot is built in the dimensions shown below. How much did it cost to built the fence?

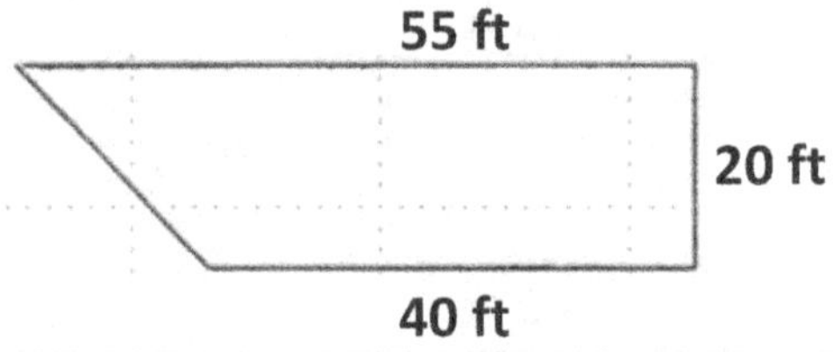

630

5) Jackson Middle arranged for 75 students to go to the theater to see a play. Some went in cars and some on a bus. Afterwards, the same bus and cars brought them back. Going to the play 51 students rode on the bus and 4 students rode in each car. Coming back 5 students rode in each car. How many students came back by bus?

45

6) A club of 10 people want to choose a president, vice-president, and treasurer. How many ways can they select them from the 10 people?

720

PERENNIAL MATH

TEST # 1

November 2019 Advanced Level Solutions

1) Solve for x. $2x - 5 = 13$.

$2x - 5 = 13$

$2x - 5 + 5 = 13 + 5$

$2x = 18$

$x = 9$

9

2) The local pet store sells one cat for every 3 dogs. The shop also sells 4 cats for every 6 birds. How many dogs does the store sell for every bird? The ratio of cats to dogs is 1:3. The ratio of cats to birds is 4:6, which is also 2:3. Therefore, since the ratio of cats to dogs is 1:3, this is also 2:6. So, if 2 cats are equal to six dogs and also equal to three birds, then 6 dogs is equal to three birds. 6:3 is equal to 2.

2

3) Charmin ordered a 4 ft. long submarine sandwich to bring to the picnic. She made 15 cuts equally spaced along the length. How many inches long was each piece? First convert the sandwich length to inches. 4ft = 48 inches. Now, if she made 15 cuts, there would be 16 pieces of sandwich. Divide 48 inches into 16 equal parts. 48/16 = 3 inches.

3

PERENNIAL MATH

TEST #1

November 2019 Advanced Level Solutions

4) A fence that costs $4.50 per foot is built in the dimensions shown below. How much did it cost to built the fence? First solve for the length of the fence by solving for the unknown side of the triangle. Use the Pythagorean Theory, $a^2 + b^2 = c^2$. Let a = 55 – 40 = 15, and b = 20. So, $a^2 + b^2 = c^2$ is $15^2 + 20^2 = c^2$, 225 + 400 625. c = 25. Now you can add the sides together: 55 ft + 20 ft + 40 ft + 25 ft = 140 ft. The cost of the fence is $4.50 x 140 = $630.

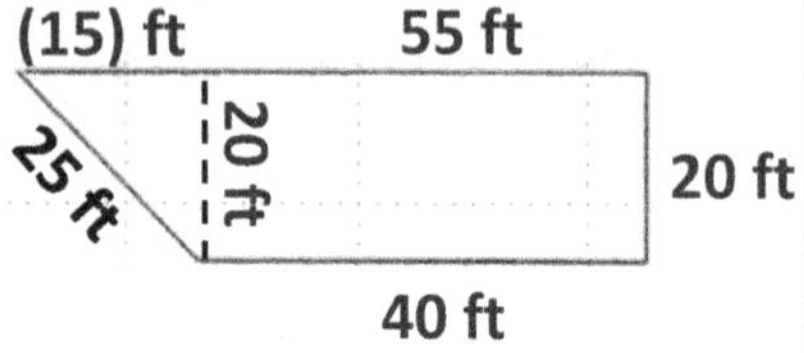

630

5) Jackson Middle arranged for 75 students to go to the theater to see a play. Some went in cars and some on a bus. Afterwards, the same bus and cars brought them back. Going to the play 51 students rode on the bus and 4 students rode in each car. Coming back 5 students rode in each car. How many students came back by bus? First solve for the number of students going by car: 75 – 51 = 24 students going by car. That means 24/4 = 6 cars. If 5 students rode in the 6 cars coming back, that would be 6 cars x 5 students = 30 students. 75 – 30 students = 45 students riding back in the bus.

45

6) A club of 10 people want to choose a president, vice-president, and treasurer. How many ways can they select them from the 10 people? If we assume that one person can only hold one office, then the first office would have 10 people to select from. After that the next office would have 9 choices, and then the final position would have 8 choices. 10 x 9 x 8 = 720 ways to select the officers.

720

Name________________________________

PERENNIAL MATH

TEST #2

December 2019 Advanced Level

1) Find the slope of the line containing the points (9, 6) and (17, 14). *(Answer must be simplified/reduced)*	
2) The ratio of the angles in a triangle is 5:6:7. What is the product of the two largest angles?	
3) It takes Larry 20 hours to complete a project alone. It takes Brett 12 hours to complete the same project. How many hours would it take them if they work together?	
4) A circle centered at A with a radius of 6 overlaps a square as shown in the picture. What is the area of the shaded region? (Use 3.14 for π) 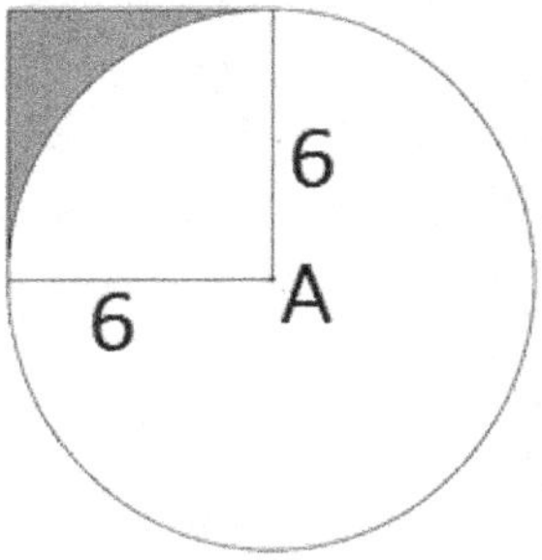	
5) A lacrosse team has won 20 games and lost 15 games. If the team wins all of its remaining games, they will have won 80% of the games they played. How many games will they play altogether?	
6) The arithmetic mean of five positive integers is 40. What is the greatest possible value of their median if only 2 numbers can be the same?	

PERENNIAL MATH TEST #2

December 2019 Advanced Level Answers

1) Find the slope of the line containing the points (9, 6) and (17, 14). *(Answer must be simplified/reduced)*	1
2) The ratio of the angles in a triangle is 5:6:7. What is the product of the two largest angles?	4200
3) It takes Larry 20 hours to complete a project alone. It takes Brett 12 hours to complete the same project. How many hours would it take them if they work together?	7.5
4) A circle centered at A with a radius of 6 overlaps a square as shown in the picture. What is the area of the shaded region? (Use 3.14 for π) 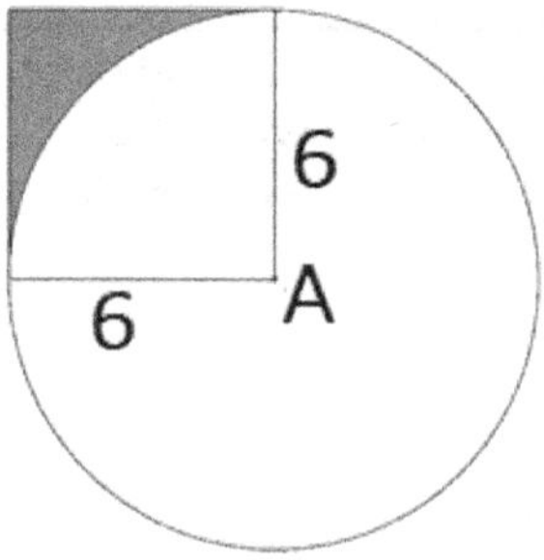	7.74
5) A lacrosse team has won 20 games and lost 15 games. If the team wins all of its remaining games, they will have won 80% of the games they played. How many games will they play altogether?	75
6) The arithmetic mean of five positive integers is 40. What is the greatest possible value of their median if only 2 numbers can be the same?	65

PERENNIAL MATH

TEST # 2

December 2019 Advanced Level Solutions

1) Find the slope of the line containing the points (9, 6) and (17, 14). The formula for finding the slope of a line is $\frac{y_2 - y_1}{x_2 - x_1}$

$$\frac{14-6}{17-9} = \frac{8}{8} = 1$$

1

2) The ratio of the angles in a triangle is 5:6:7. What is the product of the two largest angles? First divide 180° by 18 portions since 5 + 6 + 7 is 18 portions. Each portion is 10°. Therefore the angles of the triangle are 50°:60°:70°. The product of the two largest angles is 60 x 70 = 4200.

4200

3) It takes Larry 20 hours to complete a project alone. It takes Brett 12 hours to complete the same project. How many hours would it take them if they work together? To find out how much they can do together per hour add together what they can do individually per hour. So, per hour, their labors are:

$$\frac{1}{20} + \frac{1}{12} = \frac{3}{60} + \frac{5}{60} = \frac{8}{60} = \frac{2}{15}$$

Together they can do $\frac{2}{15}$ job per hour. Now divide by the time to find out how much they can do together. So now I'll pick the variable "t" to stand for how long they take to do the job together. Then they can do: $\frac{1}{t}$

$$\frac{1}{t} = \frac{2}{15}$$

$$\frac{t}{1} = \frac{15}{2} = 7.5 \text{ hours}$$

7.5

PERENNIAL MATH TEST #2

December 2019 Advanced Level Solutions

4) A circle centered at A with a radius of 7 overlaps a square as shown in the picture. What is the area of the shaded region? *(Use 3.14 for π)* First find the area of the square and then subtract ¼ the area of the circle. The area of the square is 6 x 6 = 36 units². The area of the circle is $\pi (r^2) = \pi (6^2) = 3.14 \times 36 =$ 113.04 units². Now to find the area of ¼ of the circle, divide 113.04 units² /4 = 28.26 units². 36 units² – 28.26 units² = 7.74 units².

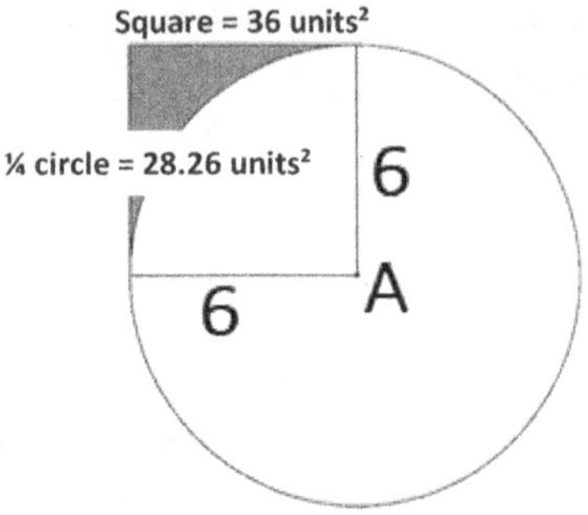

7.74

5) A lacrosse team has won 20 games and lost 15 games. If the team wins all of its remaining games, they will have won 80% of the games they played. How many games will they play altogether? Since the team will win all the remaining games, then 15 losses is equal to 20%. Then 15/2 = 7.5 games is 10%. So, 80% of the games is equal to 7.5 x 8 = 60 games. The total number of games is 60 (wins) + 15 (losses) = 75 total games.

75

6) The arithmetic mean of five positive integers is 40. What is the greatest possible value of their median if only 2 numbers can be the same? The least possible values of the first 2 integers are 1 and 1. The other 3 numbers must be nearly equal to each other to get the greatest median, but they cannot be the same. The mean of the five integers is 40 so their sum is 200. To find the sum of the remaining three integers, subtract 200 - 2 = 198. The average of the last 3 integers is 198/3 = 66. So, the last three integers are 65, 66, and 67. Therefore, the greatest possible value of the median of all 5 positive integers (1, 1, <u>65</u>, 66, 67) is 65.

65

Name________________________________

PERENNIAL MATH

TEST #3

January 2020 Advanced Level

1) What is 20% of 40% of 300?

2) Jeff has ten coins total made up of quarters, dimes and nickels. If he has at least one of each coin, how many different possible amounts could he make?

3) A bank vault is in the shape of a rectangular prism. The dimensions of the floor is 27 feet by 4 feet. The volume of the vault is 1728 cubic feet. What is the height of the vault?

4) Using the five digits 1, 2, 4, 5 and 7, how many four-digit odd numbers can be formed, with no repeated digits?

5) A total of 464 tickets were sold for the basketball game. They were either adult tickets or student tickets. The number of student tickets sold was three times the number of adult tickets. How many student tickets were sold?

6) Find the area of the shaded region.

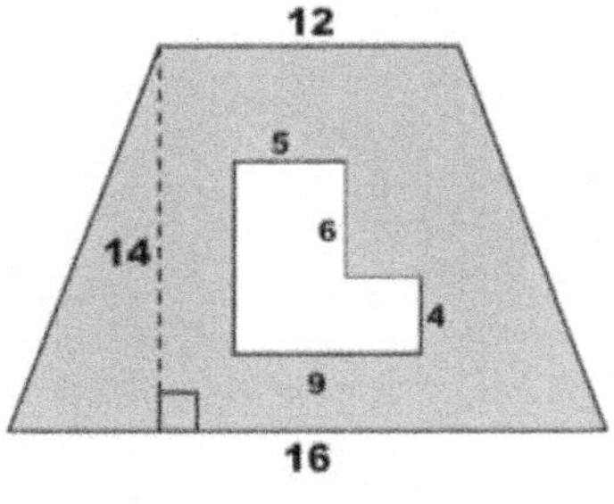

PERENNIAL MATH

TEST #3

January 2020 Advanced Level Answers

1) What is 20% of 40% of 300?	24
2) Jeff has ten coins total made up of quarters, dimes and nickels. If he has at least one of each coin, how many different possible amounts could he make?	36
3) A bank vault is in the shape of a rectangular prism. The dimensions of the floor is 27 feet by 4 feet. The volume of the vault is 1728 cubic feet. What is the height of the vault?	16
4) Using the five digits 1, 2, 4, 5 and 7, how many four-digit odd numbers can be formed, with no repeated digits?	72
5) A total of 464 tickets were sold for the basketball game. They were either adult tickets or student tickets. The number of student tickets sold was three times the number of adult tickets. How many student tickets were sold?	348
6) Find the area of the shaded region. 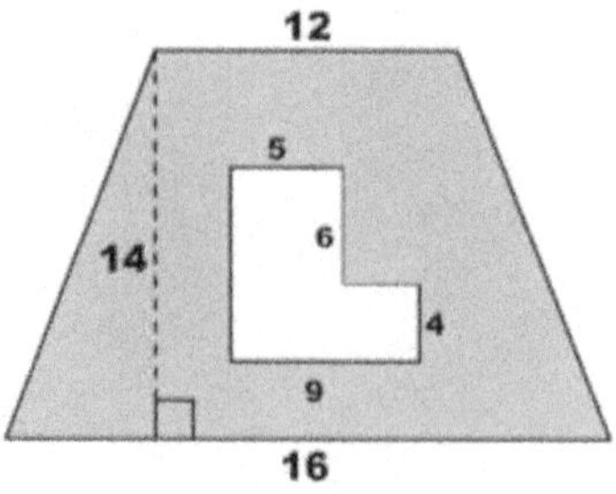	130

PERENNIAL MATH

TEST # 3

January 2020 Advanced Level Solutions

1) What is 20% of 40% of 300?
2) 300 x .4 = 120
3) 120 x .2 = 24

24

2) Make a chart to find the different amounts. There are 36 possible amounts.

Q	D	N	Total
8	1	1	2.15
7	2	1	2
7	1	2	1.95
6	3	1	1.85
6	2	2	1.8
6	1	3	1.75
5	4	1	1.7
5	3	2	1.65
5	2	3	1.6
5	1	4	1.55
4	5	1	1.55
4	4	2	1.5
4	3	3	1.45
4	2	4	1.4
4	1	5	1.35
3	6	1	1.4
3	5	2	1.35
3	4	3	1.3

Q	D	N	Total
3	3	4	1.25
3	2	5	1.2
3	1	6	1.15
2	7	1	1.25
2	6	2	1.2
2	5	3	1.15
2	4	4	1.1
2	3	5	1.05
2	2	6	1
2	1	7	0.95
1	8	1	1.1
1	7	1	1
1	6	2	0.95
1	5	3	0.9
1	4	4	0.85
1	3	5	0.8
1	2	6	0.75
1	1	7	0.7

36

3) A bank vault is in the shape of a rectangular prism. The dimensions of the floor is 27 feet by 4 feet. The volume of the vault is 1728 cubic feet. What is the height of the vault? First find the area of the floor. 27 ft x 4 ft = 108 ft^2. Now divide to find the dimension of the height. 1728/108 = 16 ft.

16

PERENNIAL MATH TEST #3

January 2020 Advanced Level Solutions

4) Using the five digits 1, 2, 4, 5 and 7, how many four-digit odd numbers can be formed, with no repeated digits? The ones digit must be odd and is thus restricted to 3 possibilities, 1, 5, or 7. Remaining digit places (thousands, hundreds, tens) are then restricted to 4, 3, 2 digit choices, respectively. There are 3 x 4 x 3 x 2 = 72 distinct permutations of 4 digit odd numbers possible to be formed from the digits 1, 2, 4, 5, and 7.

72

5) A total of 464 tickets were sold for the basketball game. They were either adult tickets or student tickets. The number of student tickets sold was three times the number of adult tickets. How many student tickets were sold? Use algebra.

$x+3x = 464$

$4x = 464$

$x = 116$ adults

The number of student tickets is 3x or 3(116) = 348.

348

6) To find the area of the shaded region find the area of the trapezoid then subtract the area of the irregular shape. The area of the trapezoid is $\frac{a+b}{2}h$.

$A = \frac{12+16}{2} \cdot 14 = 14 \cdot 14 = 196.$ Now find the area of the irregular shape inside the trapezoid. The shape is made up of 2 shapes: a 5 x 10 rectangle and 4 x 4 square is one way to break the shape up. So the area of the shape is 50 + 16 = 66. Subtract this from the area of the trapezoid. 196 – 66 = 130.

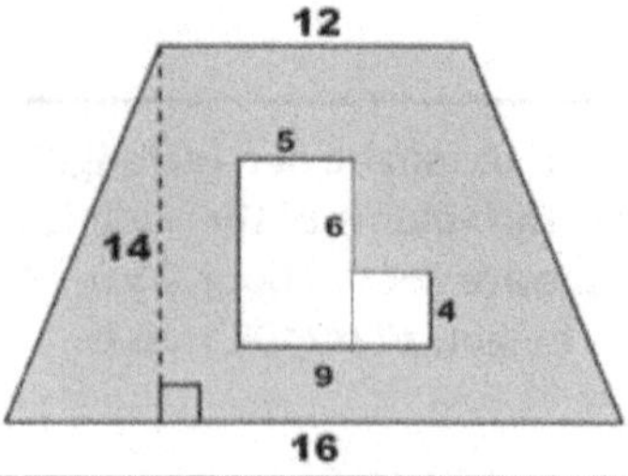

130

Name________________________________

PERENNIAL MATH

TEST #4

February 2020 Advanced Level

1) How many minutes are in February if it isn't a leap year?	
2) Kristen is three years older than Sylvia. If the product of Kristen's age in four years from now and Sylvia's age three years ago is 75, how old is Kristen now?	
3) If a whole number N is divided by 19, the quotient is 47, and the remainder is 16. What is the sum of the digits of N?	
4) There are 95 students at Davis Middle School, where the ratio of girls to boys is 11 to 8. At Johnson Middle School, there are 130 students and the ratio of girls to boys is 8 to 5. If the schools were combined into one, what would be the ratio of girls to boys? Express your answer as a simplified ratio in the form *a*:*b*	
5) In the following cryptarithm, different letters represent different non-zero digits. What digit does P represent? PQRSTSQ QRSTSQ RSTSQ STSQ TSQ SQ + Q PPPPPPP	
6) What is the 45th term of this arithmetic sequence: 9, 15, 21, 27…?	

PERENNIAL MATH

TEST # 4

February 2020 Advanced Level Answers

1) How many minutes are in February if it isn't a leap year?	40320
2) Kristen is three years older than Sylvia. If the product of Kristen's age in four years from now and Sylvia's age three years ago is 75, how old is Kristen now?	11
3) If a whole number N is divided by 19, the quotient is 47, and the remainder is 16. What is the sum of the digits of N?	18
4) There are 95 students at Davis Middle School, where the ratio of girls to boys is 11 to 8. At Johnson Middle School, there are 130 students and the ratio of girls to boys is 8 to 5. If the schools were combined into one, what would be the ratio of girls to boys? Express your answer as a simplified ratio in the form *a*:*b*	3:2
5) In the following cryptarithm, different letters represent different non-zero digits. What digit does P represent? PQRSTSQ QRSTSQ RSTSQ STSQ TSQ SQ + Q PPPPPPP	8
6) What is the 45th term of this arithmetic sequence: 9, 15, 21, 27...?	273

PERENNIAL MATH

TEST #4

February 2020 Advanced Level Solutions

1) How many minutes are in February if it isn't a leap year? First solve for how many minutes are in a day and then multiply that by how many days are in February. 60 minutes x 24 hours = 1440 minutes per day. 1440 minutes x 28 days = 40,320 minutes in February.

40,320

2) Kristen is three years older than Sylvia. If the product of Kristen's age in four years from now and Sylvia's age three years ago is 75, how old is Kristen now?

Let x = Sylvia's present age; x + 3 = Kristen's present age; x + 3 + 4 = x + 7 = Kristen's age 4 years from now; x - 3 = Sylvia's age 3 years ago.

$(x + 7)(x - 3) = 75$

$x^2 + 4x - 21 = 75$

$x^2 + 4x - 96 = 0$

$(x - 8)(x + 12) = 0$

$x = -12$ (reject, $x>0$)

or

$x = 8$

$x + 3 = 11$

Sylvia's present age = 8

Kristen's present age = 11

11

3) If a whole number N is divided by 19, the quotient is 47, and the remainder is 16. What is the sum of the digits of N?

Work backwards.

19 x 47 = 893

893 + 16 = 909

N = 909

The sum of the digits of 909 = 18

18

PERENNIAL MATH

TEST # 4

February 2020 Advanced Level Solutions

4) At Davis Middle School, the 11 parts girls and 8 parts boys make 19 parts in all, so there must be 95 ÷ 19 = 5 students in each part. That's 11 x 5 = 55 girls and 8 x 5 = 40 boys. At Johnson Middle School, the 8 parts girls and 5 parts boys is 13 parts in all, so there must be 130 ÷ 13 = 10 students in each part. That's 8 x 10 = 80 girls and 5 x 10 = 50 boys. If the schools were combined, there would be 55 + 80 = 135 girls and 40 + 50 = 90 boys. These numbers are both multiples of 45, so we get a girl to boy ratio of 135:90 = 3:2.

3:2

5) In the following cryptarithm, different letters represent different non-zero digits. What digit does P represent? First explore with the value of Q. Q cannot be 0 or 5. Use trial and error. If Q is 2, P has to be 4. This does not work out in the tens digit when you carry the 1. Q = 4. P represents the digit 8.

PQRSTSQ	8426964
QRSTSQ	426964
RSTSQ	26964
STSQ	6964
TSQ	964
SQ	64
+ Q	+ 4
PPPPPPP	8888888

8

6) What is the 45th term of this arithmetic sequence: 9, 15, 21, 27...?

The formula of an arithmetic sequence is $a_n=a_1+d(n-1)$. The common difference d is 6. Also, n represents the term in the sequence.

Substitute in the values of $a_1 = 9$ and $d = 6$

$a_n = 9+(6)(n-1)$

$a_{45} = 9+(6)(45-1)$

Simplify each term.

$a_{45} = 9+(6)(44) = 273$

The 45th term of this arithmetic sequence is 273.

273